FAST AND SLOW ANIMALS

MAMMALS

BY ERIC GERON

Children's Press
An imprint of Scholastic Inc.

THREE-TOED SLOTH

A special thank-you to the team at the Cincinnati Zoo & Botanical Garden for their expert consultation.

Copyright © 2023 by Scholastic Inc.

Library of Congress Cataloging-in-Publication Data Available
Identifiers: LCCN 2022002764 (print)
ISBN 9781338836585 (library binding) | ISBN 9781338836592 (paperback)

10 9 8 7 6 5 4 3 2 1   23 24 25 26 27

Printed in China   62
First edition, 2023

Book design by Kay Petronio

Photos ©: cover top, back cover center left, 1 top, 2 right: MerlinTuttle.org/Science Source; 2 right: Karine Aigner/NPL/Minden Pictures; 4 bottom left: Thomas Haupt/imageBROKER/age fotostock; 4 bottom right: Charlie Summers/NPL/Minden Pictures; 5 center: Kim Taylor/NPL/Minden Pictures; 6: Sean Crane/Minden Pictures; 7: Juan Carlos Vindas/Getty Images; 12-13: Heinrich vanden Berg/Gallo Images/age fotostock; 14-15: Christophe Migeon/Biosphoto; 16-17: Michael S. Nolan/Blue Planet Archive; 18-19: Jami Tarris/Getty Images; 22-23: milehightraveler/Getty Images; 24-25: J.-M Labat & F. Rouquette/Biosphoto; 26: MerlinTuttle.org/Science Source; 27: Rolf Nussbaumer/imageBROKER/Shutterstock; 28-29: MerlinTuttle.org/Science Source; 30 top left: Merlin Tuttle.org/Science Source; 30 bottom left: Stefano Ronchi/NIS/Minden Pictures; 30 bottom center: Christopher Swann/Biosphoto; 30 bottom right: HwWobbe/Getty Images. All other photos ©: Shutterstock.

MEXICAN FREE-TAILED BAT

CONTENTS

MEET the MAMMALS

Welcome to the world of mammals! This group of animals includes sloths, sea lions, bears, dogs, and porpoises, to name a few. These mammals may look very different from one other, but many have the same qualities. All mammals have hair. They are all **warm-blooded**. This means their body temperatures do not change with their surroundings.

How Mammals Move

All mammals can move! But . . . how do they do it? Cats and dogs can walk and run. Rabbits can jump and hop. Monkeys can swing from branch to branch. And dolphins swim and dive in the ocean. Mammals move in many ways! Get ready to discover how 10 mammals can travel, from the slowest to the fastest!

FACT Scientists who study mammals are called **mammalogists** (ma-MA-luh-jists).

THREE-TOED SLOTH

The three-toed sloth is the world's slowest mammal. When this furry animal is not completely still, it moves at about 0.15 miles per hour (0.24 kph). To compare, a human walks at an average speed of 2.8 miles per hour (4.5 kph). That's almost 19 times faster than a sloth!

The three-toed sloth hangs by its strong claws from tree branches. It is usually upside down as it moves bit by bit. Once a week, it travels down to the ground to go to the bathroom. The three-toed sloth is an **herbivore.** It only eats plants with very few **nutrients**. This means the sloth has very little energy to move. It saves its energy by keeping still. Sloths also keep still so **predators** cannot spot them.

THREE-TOED SLOTH CLOSE-UP

Sloths are nocturnal, which means they are active at night. They sleep 20 hours a day! Sloths are only active for about four hours at night. They tend to live alone in trees. Young sloths set up their own home when they are a few years old.

FUR
Unlike other mammals, sloths' fur grows away from their arms and legs. This helps protect their bodies from the weather.

BACK LEGS
Their back legs are very weak. Sloths rely on their front arms to move.

CLAWS
This sloth has three claws on each of its front feet. The claws help it cling to branches as it hangs upside down.

EARS

The sloth has a poor sense of hearing, so it relies on other senses like touch and smell.

FRONT ARMS

Their front arms make sloths good swimmers.

EYES

The sloth also has poor eyesight, but its eyes can see color.

SMILE

Is the sloth smiling? The fur coloring on a sloth's face is what makes it look like it is always in a good mood.

NECK

A three-toed sloth can move its head almost all the way around! That is because it has an extra bone in its neck.

#9
SQUIRREL

There are close to 300 types of squirrels. A squirrel has a bushy tail. It is a skilled climber. It scurries up and down trees by using strong claws. Squirrels gather seeds and nuts and hide them in safe places before the weather gets cold. The squirrel's sharp sense of smell helps it find where it has hidden food. A squirrel can hit the ground running at 12 miles per hour (19.3 kph).

A flying squirrel cannot fly. Instead, it has flaps of skin under its arms that look like wings and it can glide from tree to tree.

Hippos are strong swimmers and can stay underwater for 30 minutes at a time!

FACT

Hippos' skin creates an oily coating over their bodies that acts like a special sunscreen.

FACT

HIPPOPOTAMUS

The hippopotamus, commonly referred to as a hippo, is a huge mammal. It spends 16 hours a day soaking in lakes and rivers to keep cool. It leaves the water to eat grasses and plants. Hippos have stubby legs, flat feet, and webbed toes. An average female hippo weighs about 3,000 pounds (1,360.8 kg). Hippos look like they would not be able to move fast, but they can sprint across the ground at around 15 miles per hour (24.1 kph)!

#7
SEA LION

The sea lion lives in the water and on land. When it is on land, the sea lion can walk on its four flippers. In the water, the sea lion's body is streamlined and can move much faster. It can hit speeds of 25 miles per hour (40.2 kph)! Sometimes, sea lions swim in and out across the water's surface like a dolphin. This is called **porpoising**, and it helps sea lions swim even faster.

Sea lions are **carnivores**. They mostly eat squid and fish.

FACT

14

Sea lions are great at diving! They can dive down to depths of 900 feet (274.3 m) and hold their breath for 10 minutes at a time!

Out of the seven types of porpoise, the Dall's porpoise is the longest in size.

FACT

These ocean mammals have special teeth to help them hold onto slippery food like squid.

FACT

DALL'S PORPOISE

At first glance, you might think this creature swimming in the ocean is an orca, or killer whale. The Dall's porpoise also has a triangular dorsal fin, a rounded head, and a thick body. It has a white stomach and a black back. But the Dall's porpoise is smaller than an orca. A male Dall's porpoise can grow up to 8 feet (2.4 m) long. A male orca can grow up to 32 feet (9.8 m) long! The Dall's porpoise has powerful flippers and a smooth body. It can swim up to 34 miles per hour (54.7 kph).

#5
LION

The average male lion releases a burst of energy to run up to 36 miles per hour (57.9 kph). Due to their heavy weight of around 550 pounds (249.5 kg), lions cannot run for a very long time. Lions are also good at leaping. They have long tails to help their balance. They have strong muscular limbs. Lions can spend up to 21 hours a day resting. Female lions, or lionesses, do most of the hunting.

A group of lions is called a pride. Typically, a pride has 3–40 members.

FACT

FACT

A male lion's roar can be heard from 5 miles (8 km) away. That's one loud roar!

Greyhounds have been used as hunting dogs to help humans track down animals.

FACT

GREYHOUND

There are many types of dogs. The greyhound is the fastest type of dog. That is because it has strong **muscles** that allow it to run in short bursts. A fast human can run 10–15 miles per hour (16.1–24.1 kph). Greyhounds can sprint up to speeds of 43 miles per hour (69.2 kph). Their skinny bodies and long legs allow for greyhounds to run extra fast. They stretch their spines as they take lengthy strides.

#3
PRONGHORN

The pronghorn is a type of antelope. It is named after its curved horns that look like prongs. These mammals have four hooves, similar to a goat or a cow. Their front hooves are larger than their back hooves. They are able to run at 55 miles per hour (88.5 km) for up to 9 miles (14.5 km) without taking a break. The pronghorn uses its incredible speed to avoid hungry predators like wolves, cougars, and bears. They live in groups called herds.

The pronghorn is the fastest
land mammal in North America.
FACT

FACT The pronghorn's large eyes can spot
movement from 4 miles (6.4 km) away!

The cheetah has
excellent eyesight.
FACT

Cheetahs do not live together in groups
like lions. They tend to live alone.

The cheetah takes the top spot for fastest land mammal. It is known for the black spots on its yellowish coat. The cheetah is a carnivore. It eats animals like warthogs and hares. It hunts mostly during the day and sneaks up on its **prey**. Then it races forward in a burst of energy and pounces. Built for speed, a cheetah has long legs, special foot pads, and claws for stability. Its long tail helps it change direction quickly. The cheetah has been recorded running at an amazing 68 miles per hour (109.4 kph)!

#1 Fastest Mammal: MEXICAN FREE-TAILED BAT

The Mexican free-tailed bat wins for overall fastest mammal. It does not show its speed by running or swimming, but by flying. The bat is the only mammal that can fly. The Mexican free-tailed bat is the quickest mammal ever recorded. This tiny bat can fly at 98 miles per hour (157.7 kph).

Mexican free-tailed bats may be tiny, but they are super fast! They grow to be an average of 3.5 inches (8.9 cm) long and weigh 0.25 – 0.42 ounces (7.1–11.9 g). They live in large groups called colonies.

FACT Most Mexican free-tailed bats **migrate**, or fly south for the winter.

27

BODY

Their bodies are covered in dark fur that helps them blend into their habitat.

EARS

Bats use sound to hunt. They are nocturnal and hunt at night.

MOUTH

Bats make chirping and clicking sounds to communicate with each other. They have teeth to chew up bugs like ants, beetles, and flies.

TAIL

These bats are known for having long tails that trail out behind them when they fly.

WINGS

These mighty wings are the secret to the bat's speed. The bat can also use them to quickly change direction during flight.

FEET

Their feet are used for grasping onto surfaces in caves, dens, trees, and other places.

MEXICAN FREE-TAILED BAT CLOSE-UP

Mexican free-tailed bats are **insectivores**, which means they only eat insects. Predators that eat these bats include owls, hawks, snakes, skunks, and possums.

MAMMALS FAST AND SLOW

Now you know mammals can move in many ways. They run. They jump. They fly. They swim. They hang from tree branches. Some creep along slowly, but many do not. The animals in this book are only a few of the mammals in the world. There are more than 5,400 types of mammals out there. Make it your mission to learn even more about how these amazing animals move and how fast they can go!

GLOSSARY

backbone (BAK-bohn) a set of connected bones that runs down the middle of the back; also called the spine

carnivore (KAHR-nuh-vor) an animal that only eats meat

habitat (HAB-i-tat) the place where an animal or a plant is usually found

herbivore (HUR-buh-vor) an animal that only eats plants

insectivore (in-SEK-tuh-vor) an animal that only eats insects

mammalogist (ma-MA-luh-jist) a scientist who studies mammals

migrate (MY-grayt) to move from one area to another at a particular time of year

muscle (MUHS-uhl) a type of tissue in the body that can contract or produce movement

nocturnal (nahk-TUR-nuhl) active at night

nutrient (NOO-tree-uhnt) a substance such as a protein, mineral, or vitamin that is needed by people, animals, and plants to stay strong and healthy

porpoising (POR-puh-sing) coming in and out of the water while swimming at a fast rate

predator (PRED-uh-tur) an animal that lives by hunting other animals for food

prey (pray) an animal that is hunted by another animal for food

warm-blooded (WORM bluhd-id) having a body temperature that does not change, even if the temperature of the surroundings is very hot or very cold

INDEX

Page numbers in **bold** indicate images.

ABOUT THE AUTHOR

Eric Geron is the author of more than a dozen books. He lives and works in New York City. He can move faster than a three-toed sloth but sadly would not be able to outrun a hungry lion.